FaceTime® for Seniors

by

Joy Alford-Brand, JD

(c) Copyright 2020

Table of Contents:

Hi, I'm Joy Alford-Brand. My husband and I have a number of older family members who live by themselves, including my father and my in-laws. These days, we can't get together the way we used to before the Coronavirus hit. It has been a challenge staying connected with them since they aren't too familiar with using tablets, smartphones or computers.

One day, my husband decided to buy his parents an iPad®so we could video chat with them using FaceTime®. We got the iPad and set it up for them, fully expecting it to be easy for them to use FaceTime and their iPad. Well, it turned out that it was harder for them to use the iPad than we expected. Right away, I decided it was time to sit down and write a simple, step-by-step guide to using FaceTime for senior citizens.

My goal with this guide is just to take you through the basic steps you need to take to make or receive a FaceTime call. While there are lots and lots of other things you can do with your iPad, this guide is intended to do just one thing, give you a guide you can hold in your hands and use with ease.

Finally, note that pages with headings that are outlined in green are for advanced instructions. These are not necessary to FaceTime with your iPad or iPhone®, they are just fun things you can learn to do!

While this manual will help you FaceTime (or video chat with your iPad), there are some tech things you might need help with before you can get started. Only relatively recent devices will allow you to FaceTime with more than one person. You should already have iOS® 12.1. 4 or later, or iPadOS® on one of these devices: iPhone 6s or later, iPad Pro® or later, iPad Air 2® or later, iPad mini® 4 or later, iPad (5th generation) or later, or iPod touch® (7th generation). If you are not sure your device will work, you may want to ask someone to help you check to see if you can update it or if you need to get a newer device.

Your iPad should also have a WiFi connection. If it doesn't, you won't be able to FaceTime. Finally, you should have already set up your email account. If you haven't done that, you my need to ask someone for help with that, too. You should also know how to operate your iPad, meaning you know how to touch/tap, swipe and scroll on the screen with the tip of your finger.

FaceTime is an application (or program), that comes pre-installed on your apple device (your iPad or iPhone). While it is possible for you to delete it from your Home page it's pretty difficult to do. The FaceTime app icon on your device looks like this:

So, how do you make or receive calls on FaceTime? First you will need to familiarize yourself with the iPad starting with the face of your iPad and the Home Page. Did you know there are **2** cameras on your iPad, the front-facing and rear-facing cameras? To FaceTime, you will use the front-facing camera, that lets you see yourself **AND** whoever you are FaceTiming with. This is the camera that people use to take "selfie" pictures of themselves. The reason is because you can look into the camera and see the picture you are about to take on the screen at the same time.

It's also important to know where the the Home button is and what it does. The Home button is the only button on the face of your iPad. The Home button will always take you to the Home Page which is where all of your apps are located. See the picture and diagram on the next page.

This is the Face of your iPad

This is the front facing camera, small isn't it?

Power button on the top edge

Volume buttons

11:23
Saturday, July 4

FACETIME   now
Ron
To Mom & Dad, Julie, and Ron
Join FaceTime Call

try again

This is the **Wake Up** screen. It opens when you "wake up your iPad." You will get notified about missed FaceTime calls when your iPad wakes up.

This is the **Home Button**. Pressing this **once** will always get you to your Home Page.

This is the **HOME Page** of your iPad

Battery life

Time, day and date

WiFi connection. It's the same everywhere:

10:02 PM   Mon Jun 22   45%

The Calculator

Calendar
Monday
22

Photos

Camera

Clock

Apps
(applications)

Settings

App Store

You select
which apps
go on the
left side of
the bar.

Left side are your
Favorite apps

Right side are
Recently used apps

30,632

Text/Messages

FaceTime
Tap to make a
FaceTime call

Safari ®
Internet

Email

Calculator

App Store

Settings

Before you can make a FaceTime call, you will need to add people into your contacts list (electronic address book), **before** they will show up on your FaceTime contact list. We'll go over how to do that on the next page. You can add someone to a FaceTime call using their cell phone number (not their home phone number), or their email address. You will need to add one or both of those things into each person's contact information in your contacts list (or your electronic address book). Contacts list icon:

Once you've made or received a FaceTime call, that call will show up on your FaceTime home screen on the left hand side and you can simply tap on the call if you want to repeat it or call the same people again. We'll go over that on page 17.

One thing you need to know is that you **can't** FaceTime with someone who does not have an iPad or iPhone. The FaceTime app is only on iPhones, iPads and Mac®computers. There are other apps you can use to video chat with people who do not have iPhones, iPads or Mac computers.

6

This is how you add a contact to your electronic address book.

Step 1 - look for this app on your iPad. ➡️

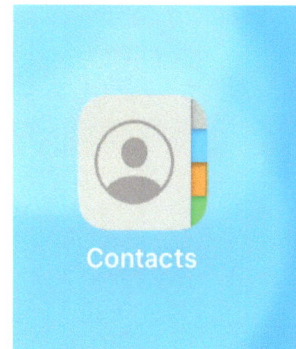

Contacts

Step 2 - tap on the app and your contact list will appear. It looks like this.

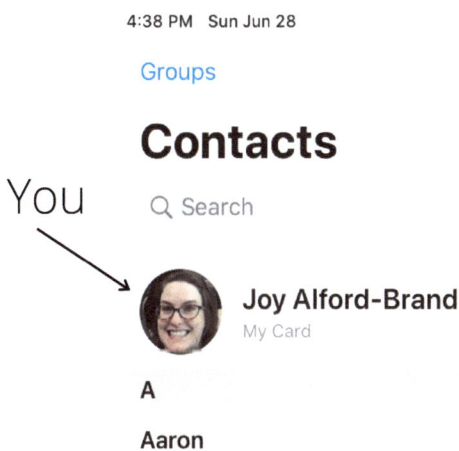

Step 3 - add a contact by tapping on the blue + sign. Then create a new contact.

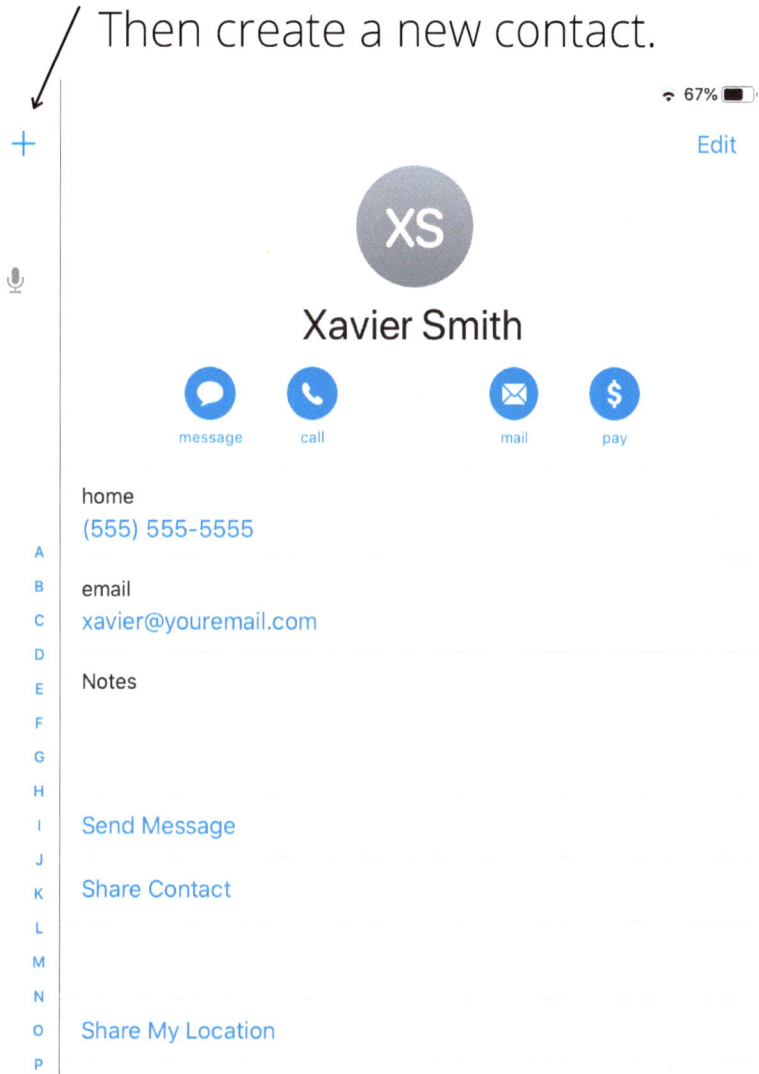

4:38 PM Sun Jun 28

Groups   +

Contacts

🔍 Search   🎤

You →

Joy Alford-Brand
My Card

A

Aaron

67% 🔋

Edit

XS

Xavier Smith

💬 message   📞 call   ✉️ mail   💲 pay

home
(555) 555-5555

email
xavier@youremail.com

Notes

A
B
C
D
E
F
G
H
I   Send Message
J
K   Share Contact
L
M
N
O   Share My Location
P

7

Step 4 - create a new contact and add contact information

You will need to add the following information for each contact to FaceTime with them.

1. Name
Just tap and type.

2. Cell Phone #
Tap the green + and type.

OR

3. Email Address
Tap the green + and type.

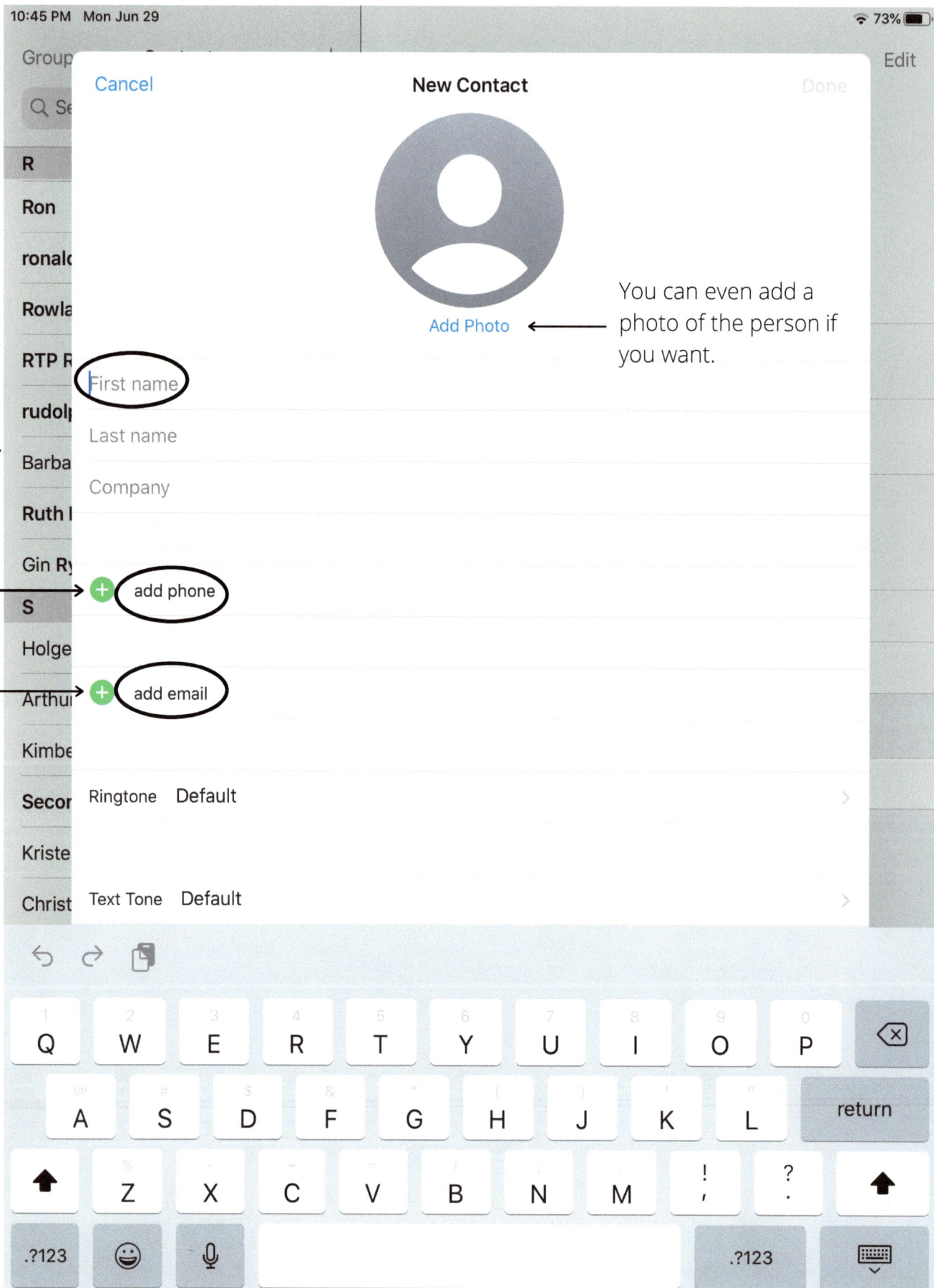

10:45 PM Mon Jun 29

Group

R

Ron

ronald

Rowla

RTP R

rudolp

Barba

Ruth

Gin Ry

S

Holge

Arthu

Kimbe

Secor

Kriste

Christ

Cancel

New Contact

Done

Add Photo ←——— You can even add a photo of the person if you want.

First name

Last name

Company

+ add phone

+ add email

Ringtone   Default   >

Text Tone   Default   >

Edit

Advanced - Adding a photo to a contact

1. Click on add photo

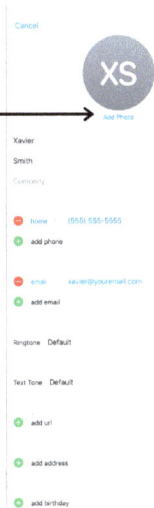

2. You have 3 options:

A. Choose from a photo in your photo folder on the device,

B. A different color for the contact icon or

C. An emoji from the emoji list.

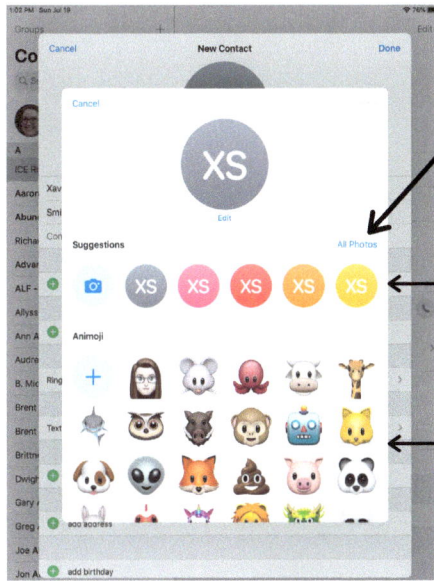

3. If you click on "All Photos," the photos you have on the iPad will pop up. Then you can pick one by tapping on it.

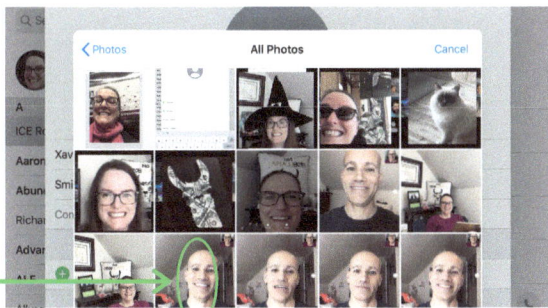

4. The photo will open in a circle. You can center the photo in the circle by dragging it with your finger.

Then tap "choose" in the lower right hand corner.

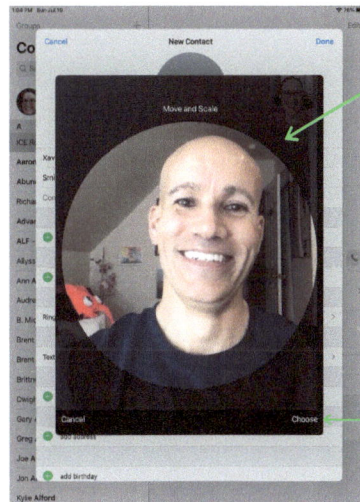

5. You can choose a filter or leave it as an original photo by clicking whichever one you want.

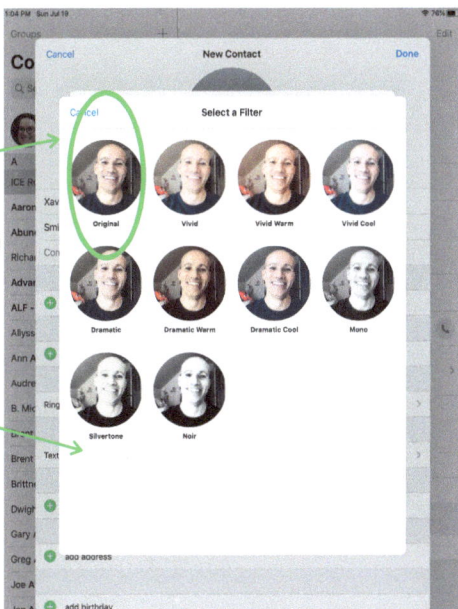

6. When the next screen pops up, click on done in the right hand corner of the white screen and you have added a photo to a contact!

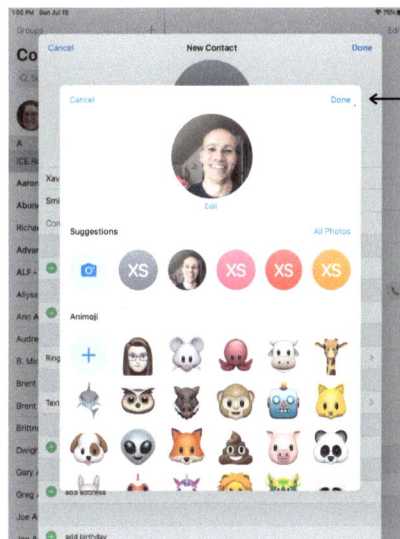

9

There are two cameras on your iPad or iPhone! The first is on the front of your iPad, we showed you where it is on page 4. The second camera is on the back of your iPad or iPhone (we are just showing you the iPad in this guide).

Camera Lens

Flash

1. To take a picture, find the camera icon on your Home Page and tap on it to open the camera. It looks like this:

2. When you tap on the camera icon, the camera opens and your screen looks like this:

Timer - sets a timer so you can be in the photo.

Flash - turns it on, off or puts it on auto.

Camera Flip - flips from front to rear camera.

Shutter Button - tap to shoot!

Preview/Camera Roll - lets you see your shot and takes you to your camera roll.

Video - tap this to film a video with your camera.

Current mode - shows the current camera mode. Photo is the default.

3. To take a picture, point the camera at the object you want to photograph and tap the shutter button. Make sure your subject is centered and not blurry and hold the device is steady. You can tap the "camera flip" button to go back and forth between the front-facing and rear-facing cameras. You would use the front-facing camera to take a selfie (a picture of yourself), and the rear-facing camera to take a photo of something else. You can also take a video from this screen by tapping the "video" button.

There are other tools you can experiment with on this screen, including the timer, flash and camera mode options. if you scroll up or down through the different modes on the camera, you can use the panorama (pano), time-lapse, slo-motion or square modes and take some really interesting photographs and videos!

Once you've taken the picture, it will be stored in your photos folder. You can find it by tapping this icon: You can also see that there are other things on this screen like your flash, timer and a small preview window that will show you what you are about to take a picture of. If you tap on the preview window it is a shortcut that will take you into your photos (also called your "camera roll").

Once you've added all the people you want to your FaceTime call, your screen will have 2 green buttons, an audio button and a video button. If you want to see **and** hear who you are FaceTiming with, tap the green video button. If you just want to call them like a regular telephone call, tap the audio button.

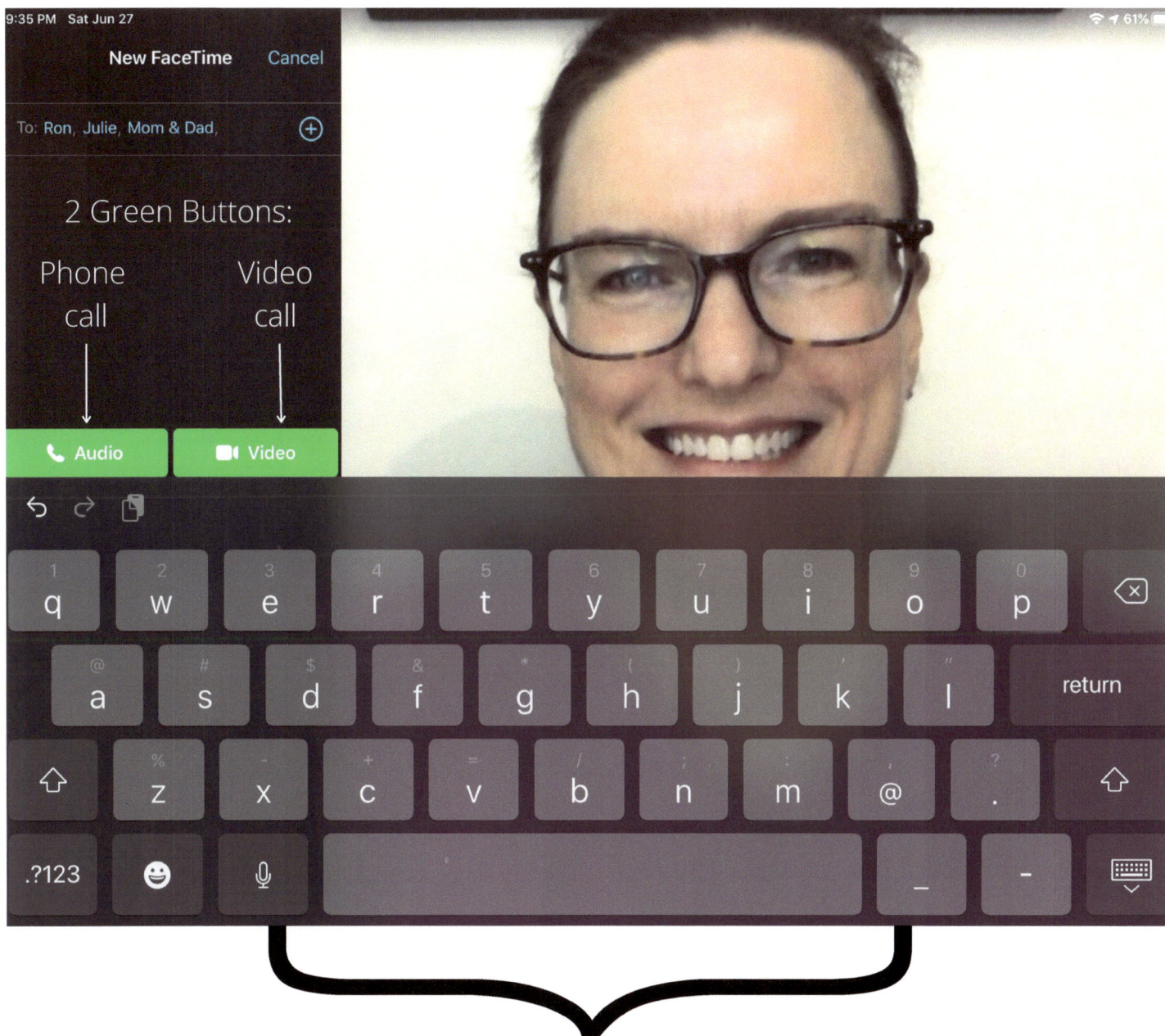

The keyboard will pop up when you are adding people to the call. It looks like a typewriter, right? Once you tap the audio or video button to make the call, the keyboard will disappear.

To make a FaceTime call, look for the app that has a green square with a white video camera on it on the homepage:

I labeled it "FaceTime" in red on page 5. Once you find that app, tap on it to open it. Once you open the FaceTime app, you will have to type in who you want to call from your contact list. Your screen will look like this and you will be able to see yourself on the right hand side.

You should see your face here, on the right side of the screen. if you can't see your face, move the iPad so you are looking into the front-facing camera (see page 1 to for front-facing camera location). If you can't see you, they can't see you either. They see what you see!

1. Type in the name of who you want to call.

2. Tap on how you want to invite them to the call. If they don't have a cell phone #, use their email Address. (You can only add names that are blue).

3. Tap the blue + to add someone else to your call.

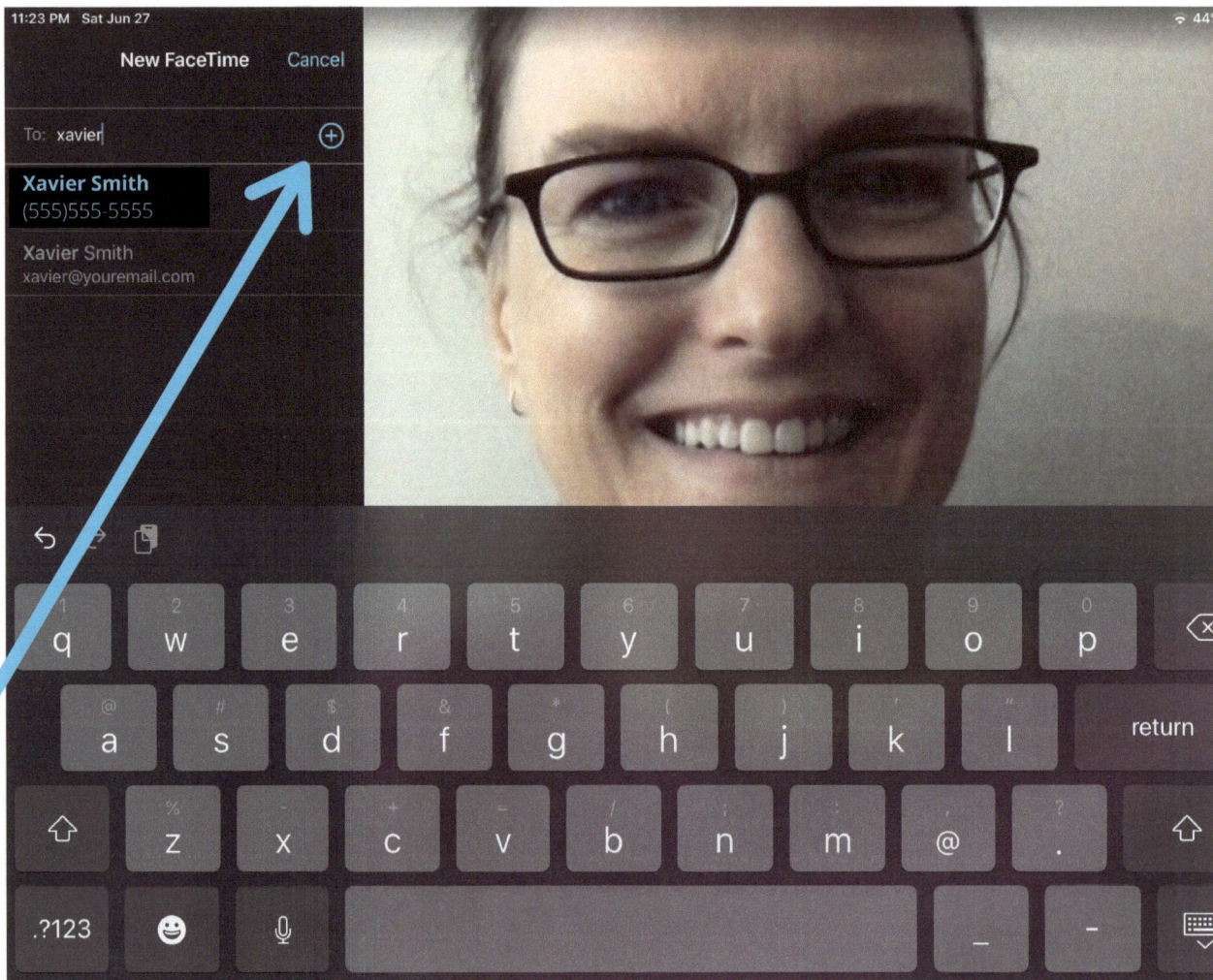

11:23 PM Sat Jun 27

New FaceTime   Cancel

To: xavier

Xavier Smith
(555)555-5555

Xavier Smith
xavier@youremail.com

4. Repeat steps 1, 2 and 3 until you've added all the people you want on this FaceTime call (up to 32 people can be added).

Adding multiple people onto the call can make family gatherings, dinners, holidays and events possible even if you can't leave your home.

12

Once you have clicked on the FaceTime call button 📷 you will have to wait until the other people on the call pick up. This is what your screen will look like while you are waiting:

Note: This is what your screen will look like with the iPad held vertically. This is called "portrait view."

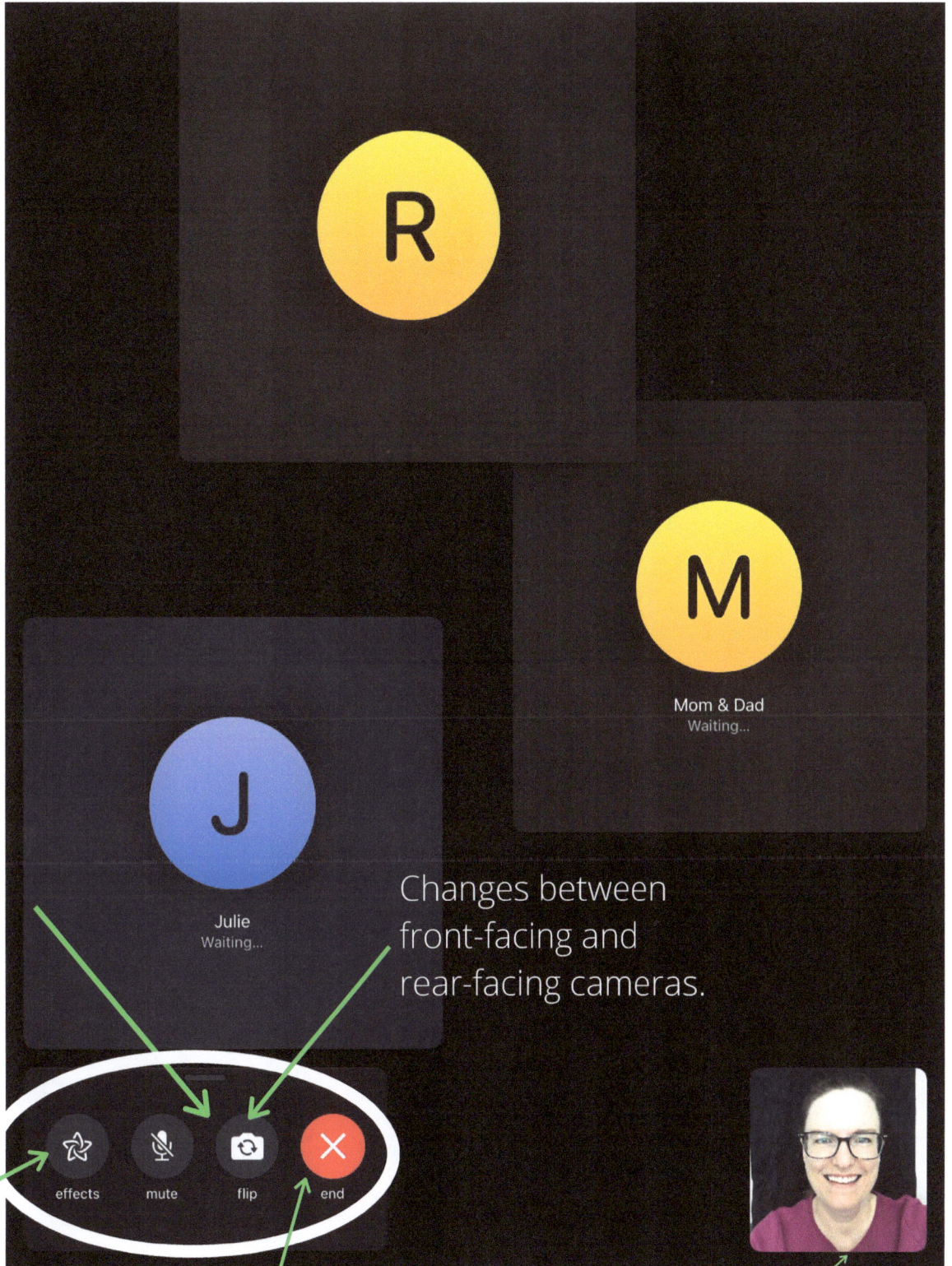

R

M

Mom & Dad
Waiting...

The 3 grey buttons are not critical to your call unless you want to mute **yourself** and just be in the background listening.

J

Julie
Waiting...

Changes between front-facing and rear-facing cameras.

effects   mute   flip   end

Allows you to add effects to your picture/video during the FaceTime call.

If you want to end the call, tap the red button with the X on it.

This is how you should see yourself somewhere on the screen, if you don't, adjust your iPad so you are looking into the front-facing camera.

13

You can add effects to your FaceTime call to make things more interesting. Here is how you use the FaceTime effects tools:

1. Tap the effects button.

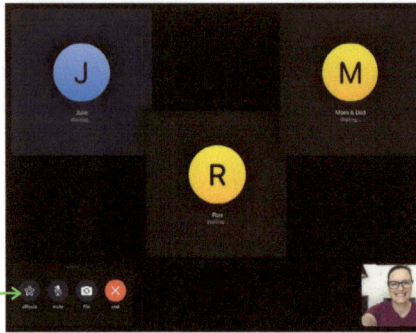

2. Choose from the effects options that pop up.

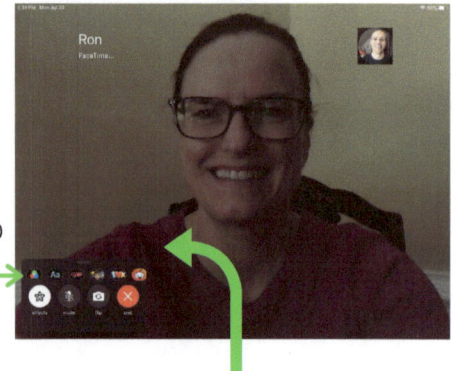

You can choose from the effects options by swiping left with your finger.

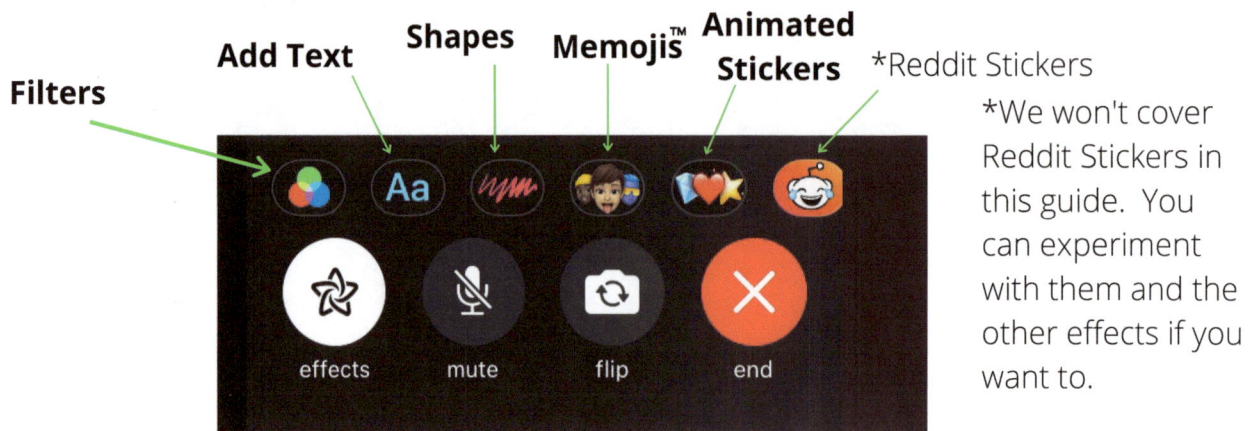

Filters

Add Text

Shapes

Memojis™

Animated Stickers

*Reddit Stickers

effects   mute   flip   end

*We won't cover Reddit Stickers in this guide. You can experiment with them and the other effects if you want to.

Filters allow you to change how your picture looks during the call. You can choose from all the options on the left hand side of the screen.

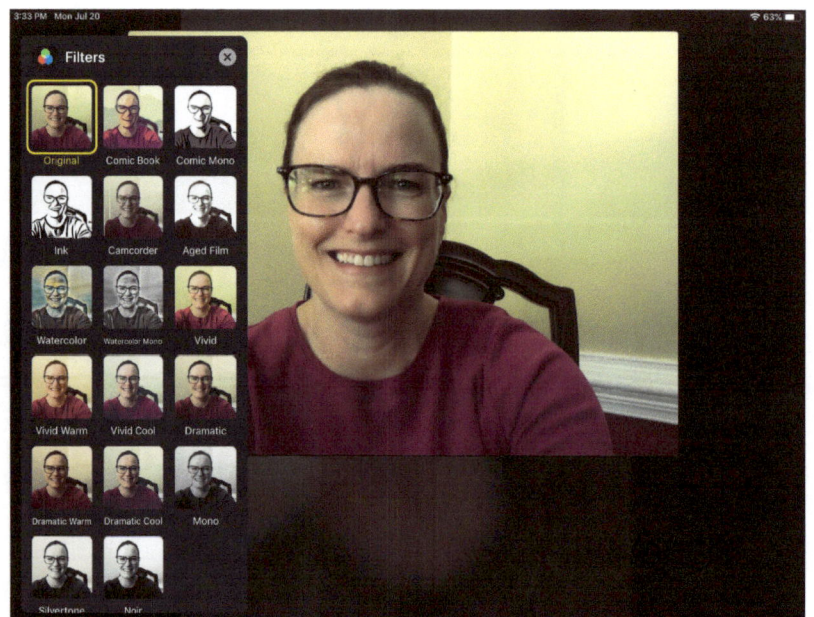

Text allows you to add text to your FaceTime call. When you tap the text button, you can choose from the types of text you want to appear on your picture during the call.

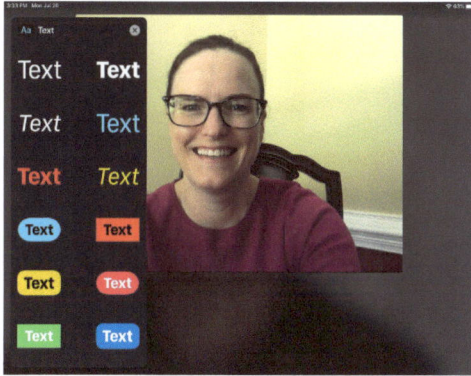

To remove the text from your picture, tap the text effect button and tap on the "X" in the corner of the text.

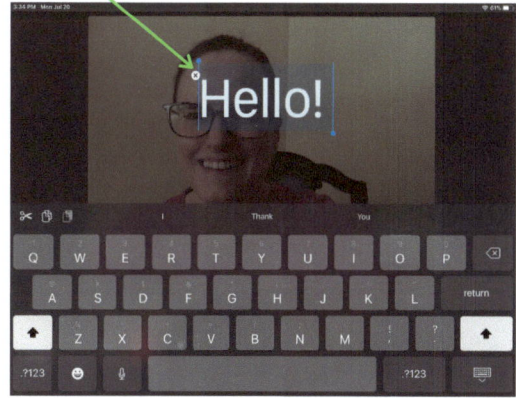

Type whatever text you want to see on your picture on the keyboard that pops up.

Shapes allow you to add images to your picture during the call. You can choose from all the options on the left hand side of the screen.

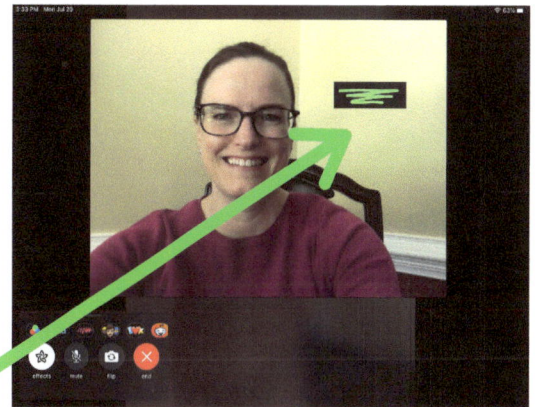

There it is!

Animated Stickers allow you to add little cartoon-type images to your picture during the call. You can choose from all the options on the left hand side of the screen.

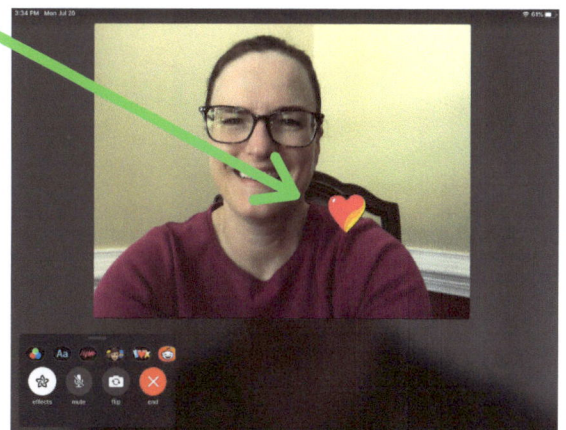

Memoji Stickers allows you to add an emoji that looks like you OR an animal from the list of options. You can choose from the list by swiping to the left.

You can set up your own memoji in the text app. It is customizable so you can make it look like you!

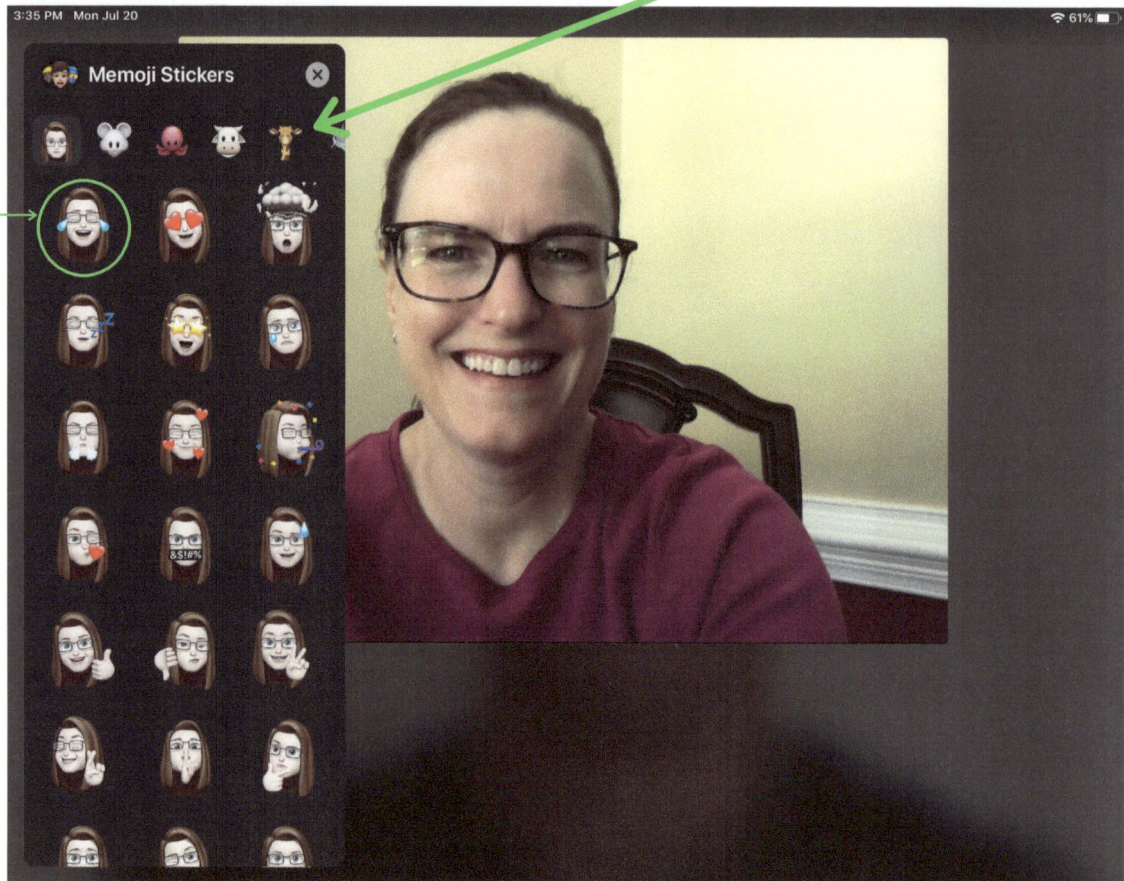

Your memoji will appear on your screen until you turn it off by tapping the effects button and bringing up the screen where you select your memoji. Then tap on the memoji and tap on the "x" which will turn it off on the picture. This works for all effects.

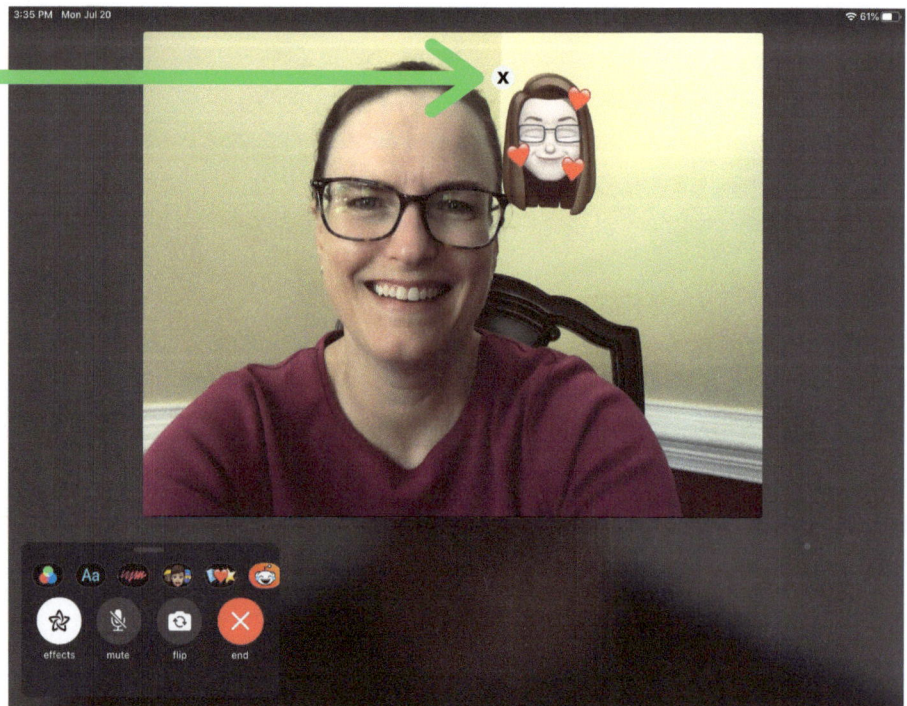

You can turn off any of the effects by tapping on the effect button you used to turn it on and tapping the "x" next to the effect. Your effect will also disappear when you end the call.

16

This is what your screen will look like when a person you called picks up and the video starts.

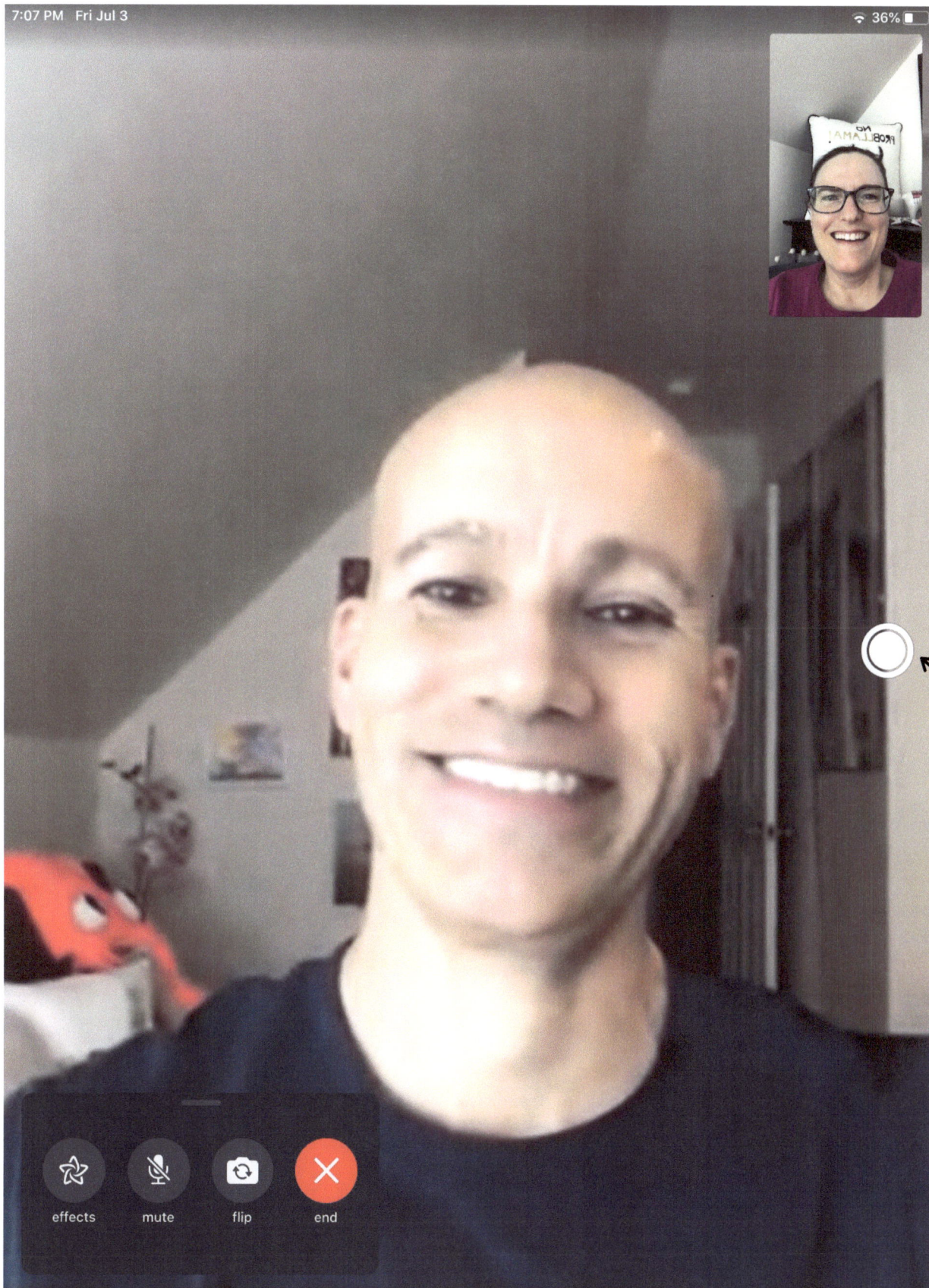

7:07 PM   Fri Jul 3

36%

You will appear in a corner of the screen during the call.

Tap this button to take a picture of your screen.

effects   mute   flip   end

There my be times when you want to take a picture of what is on the screen of your iPad. The way you take a picture of your screen (also called a screenshot), is by pressing the Power button and the Home button **AT THE SAME TIME**. When you take a screenshot, it will be saved in your "photos" app on your device. You can find it by tapping this icon:

Photos

Power button on the top right hand **edge** of the device

11:23

Saturday, July 4

FACETIME   now
Ron
To Mom & Dad, Julie, and Ron
Join FaceTime Call

try again

Home button

18

You can also turn the iPad on it's side. If you do, your screen will be longer in width, like a tv). This is called "landscape view" and it will look like this:

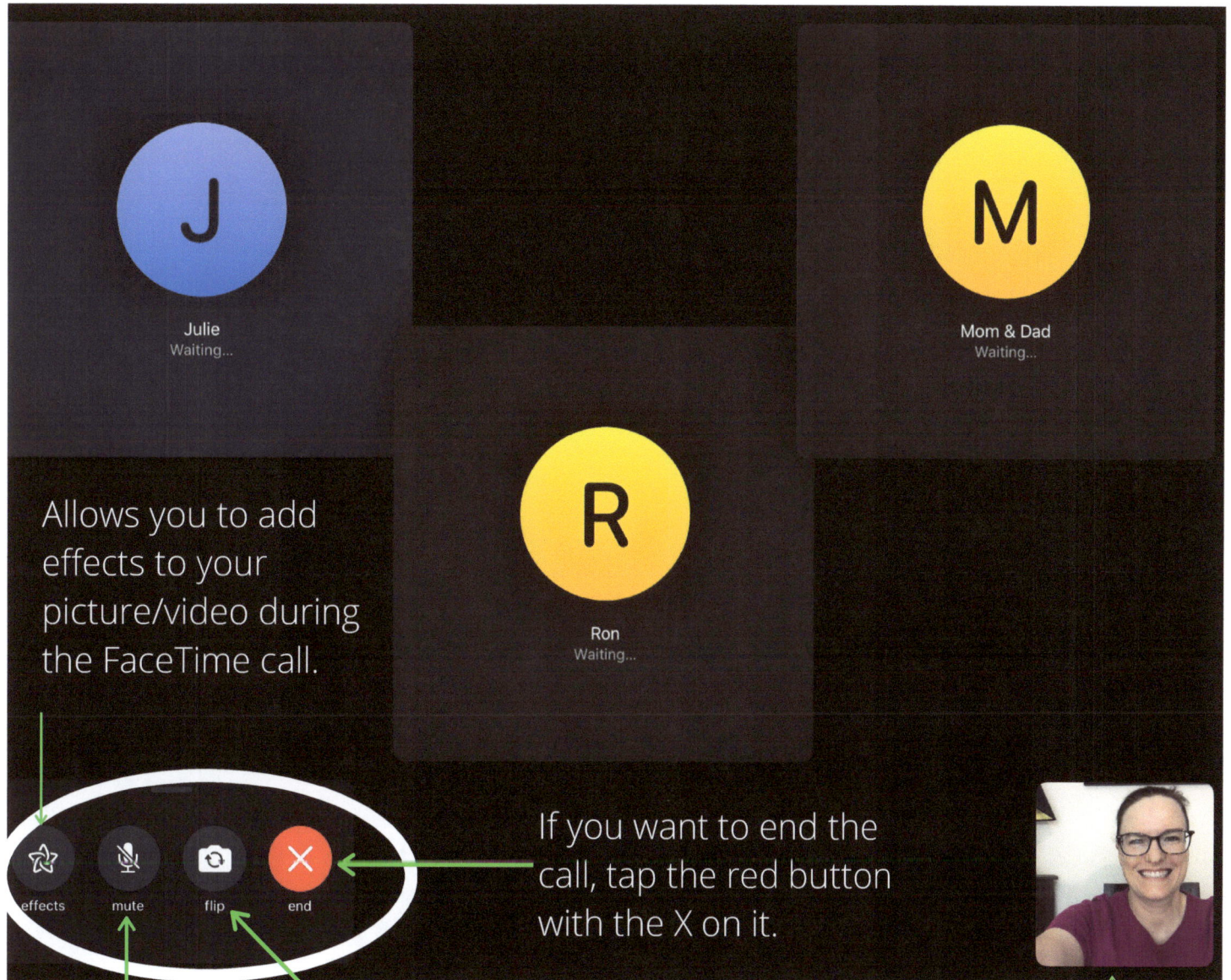

Allows you to add effects to your picture/video during the FaceTime call.

If you want to end the call, tap the red button with the X on it.

Allows you to mute **yourself** during the call if you just want to listen in.

Changes between front-facing and rear-facing cameras.

This is your picture, if you don't see yourself in it tap this icon:

Turning your iPad on it's side might be a little better to see, if there are several people on the call. The person speaking will pop out bigger during the call.

19

How do you answer a FaceTime call if someone is **calling you**?

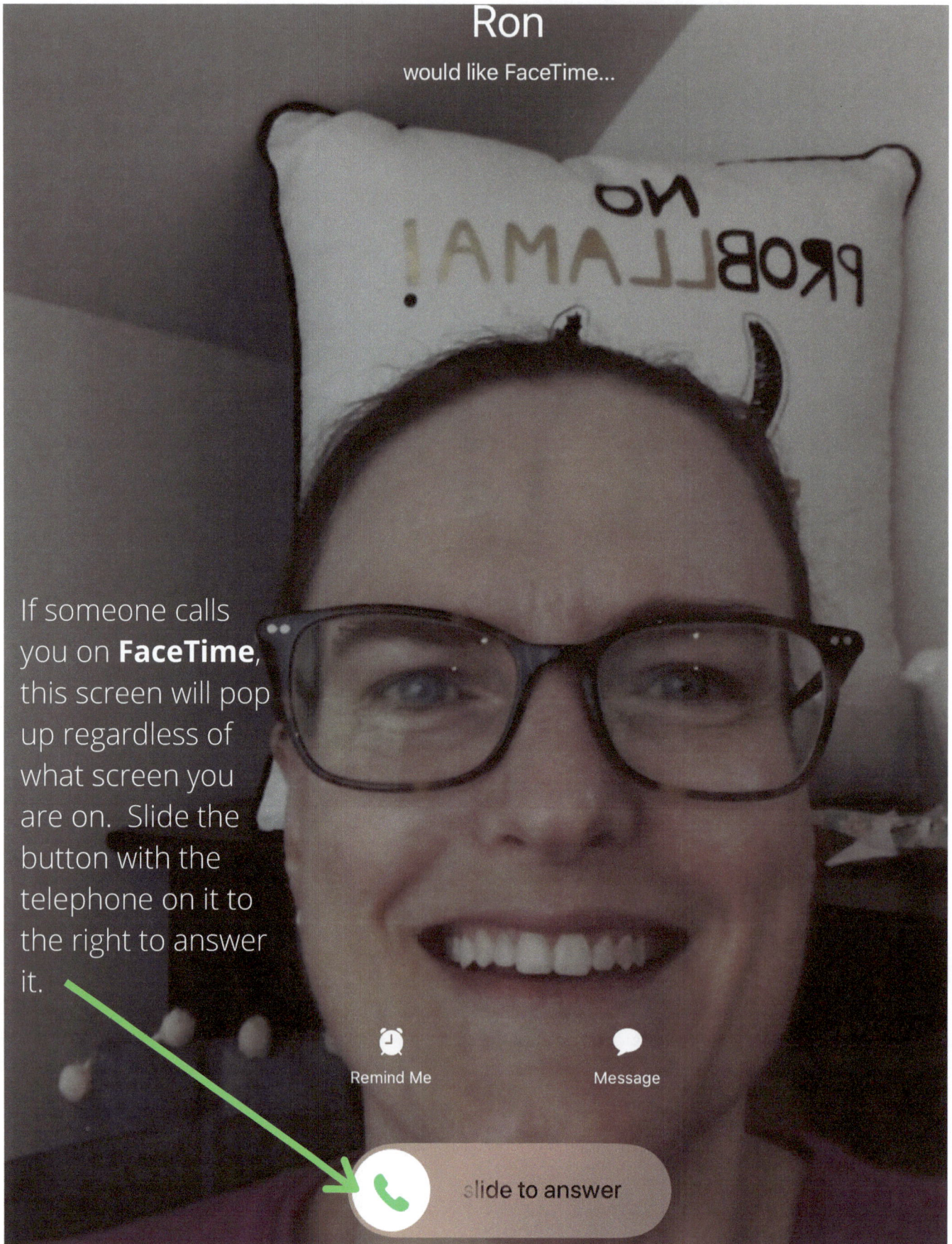

Ron

would like FaceTime...

If someone calls you on **FaceTime**, this screen will pop up regardless of what screen you are on. Slide the button with the telephone on it to the right to answer it.

Remind Me

Message

slide to answer

How do you answer a FaceTime call if someone is **calling you**?

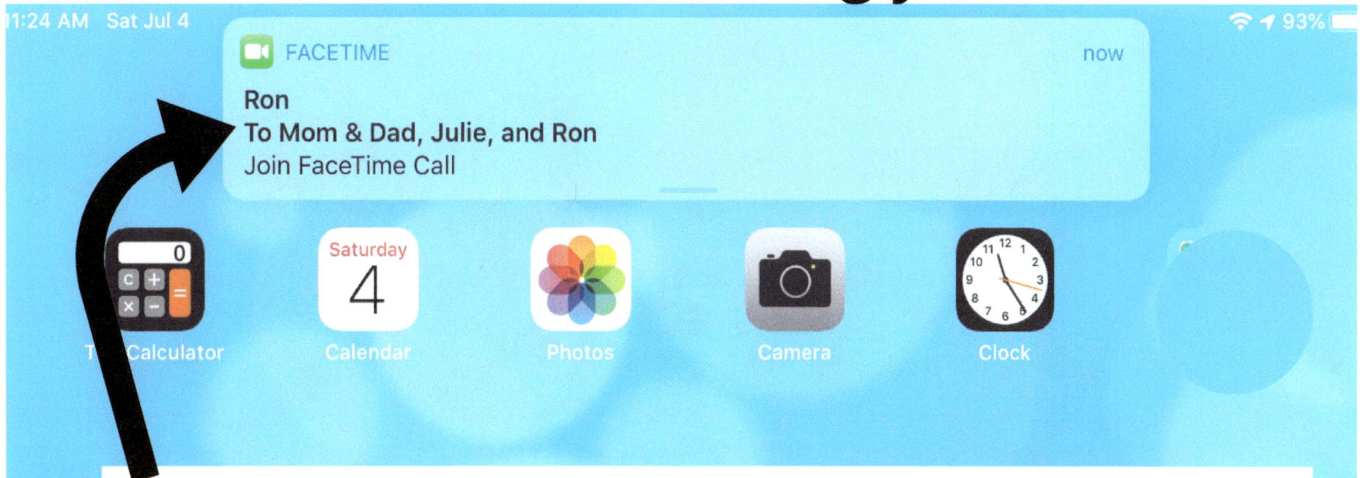

FACETIME   now

Ron
To Mom & Dad, Julie, and Ron
Join FaceTime Call

Calculator   Calendar   Photos   Camera   Clock

If someone includes you on a **group FaceTime call**, this notification bubble will pop up regardless of what screen you are on. Tap on it to answer. The pop up bubble will only last a few seconds before it disappears. If you see the notification bubble but it disappears before you can tap it, just touch the **FaceTime icon** to join the call. If the call is still in progress, it will be at the top of your call list.

FaceTime icon

30,939

21

Once you "answer" the FaceTime call you will be on this screen.
Tap the green button on the screen that says join:

9:44 PM Sat Jun 27   57%

Ron, Julie & Mom & Dad
FaceTime Call from Ron

effects   mute   flip   join

The green button
becomes the 'red, end' call
button during the call.

end

To answer the call, tap the
green button on the screen
that says join.

Recent Calls and Missed Calls

This is your recent call list.

Calls you missed will show up in red

Calls you made or received will show up in white.

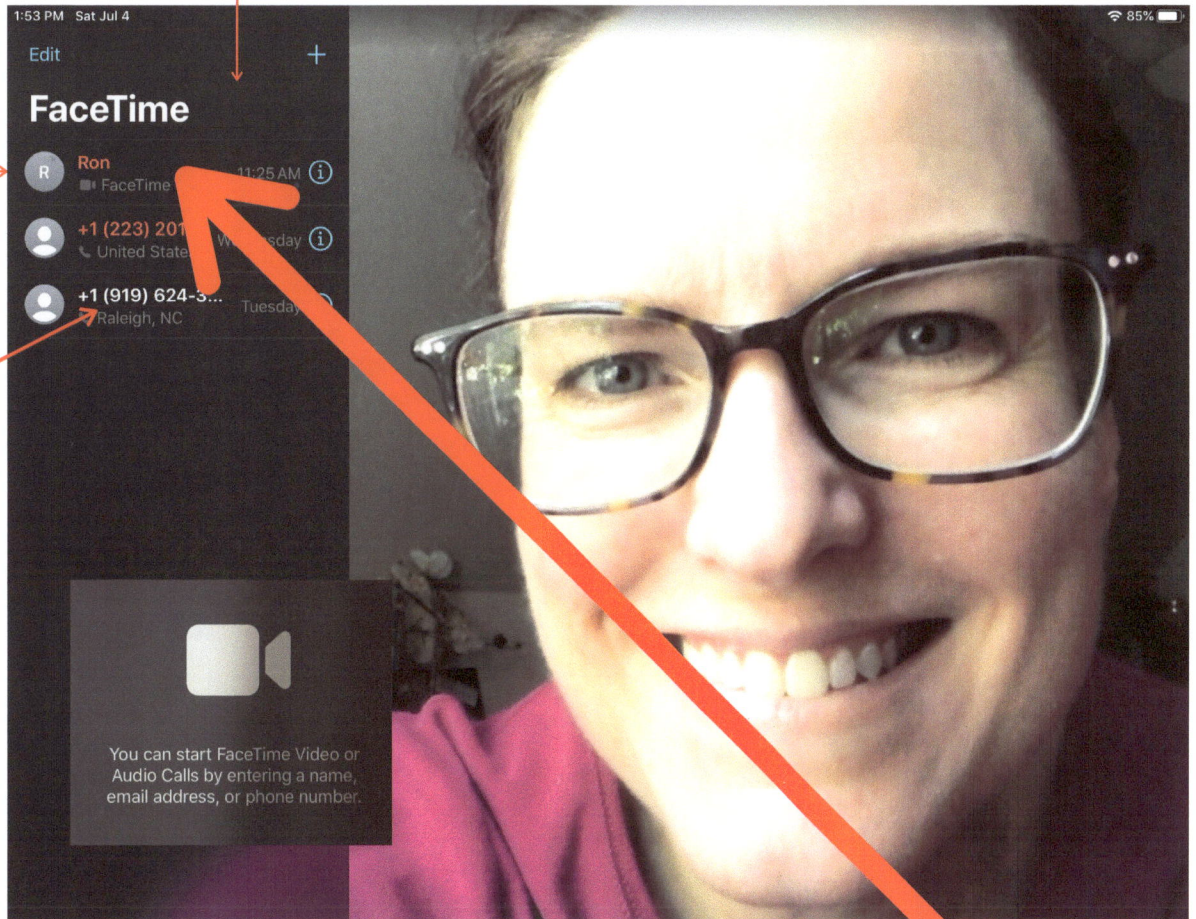

If you want to call the same people again tap on the recent call on the list you want to repeat. So if you want to call **Ron**, you would tap on his name on this list. This feature is great when you are repeating a call that has multiple people on it because you don't have to individually add each person onto the call again.

This is handy if you get together with a group on a regular basis. For example, we "get together" with our friends and family every week and it's easier to just tap on last week's call than it is to add everyone to a new call each week.

You can also tell you have missed incoming calls 2 other ways.

You will be notified that you missed FaceTime calls when your iPad wakes up. Below you can see that you missed two calls from Ron.

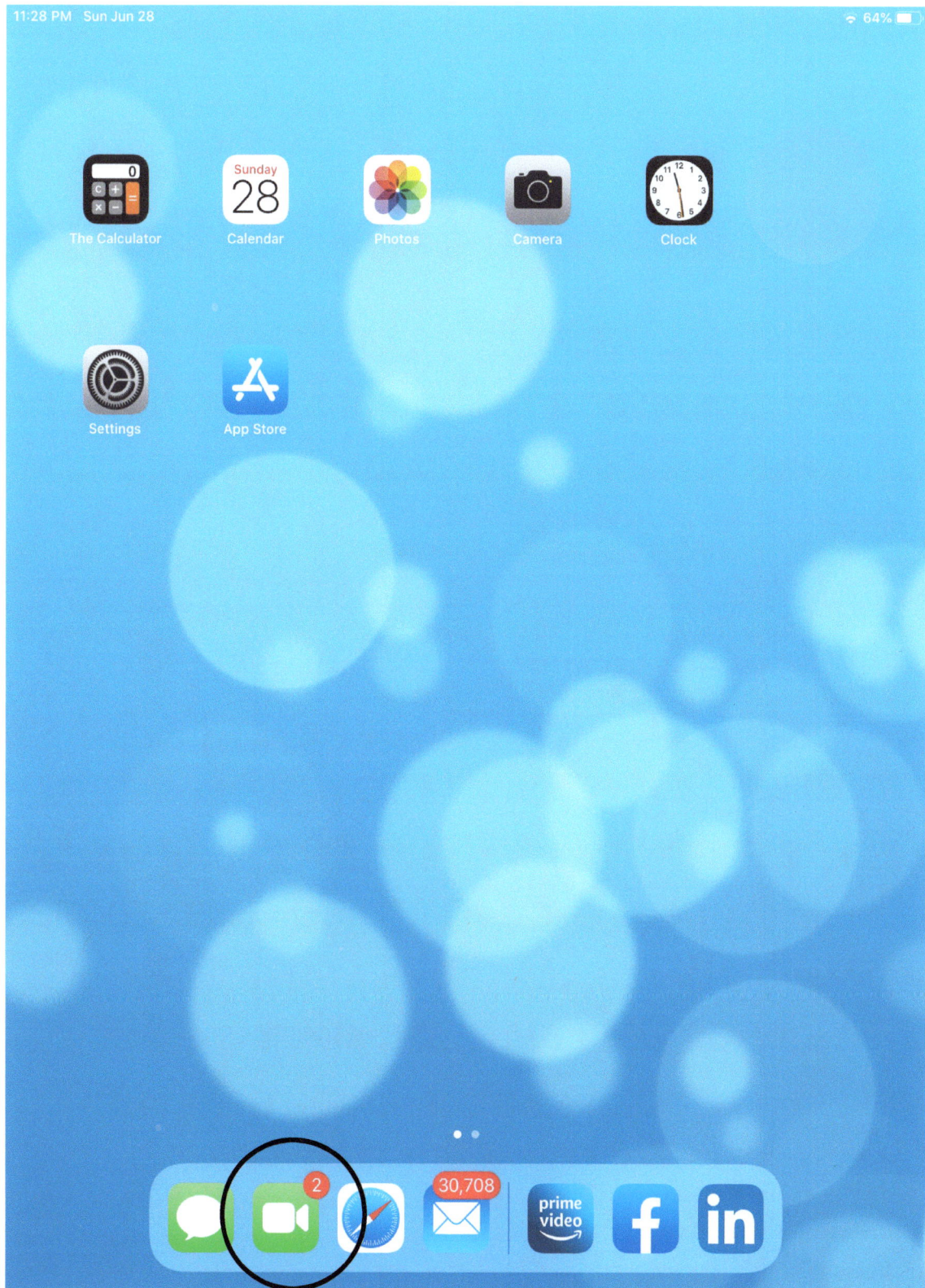

You will also be notified that you missed FaceTime calls by a number in red on the FaceTime icon on the Home Page. The number tells you how many calls you missed. Here we see we missed 2 calls.

Sometimes you could get disconnected from your FaceTime call. This is not an uncommon thing and could happen for any number of reasons. If your WiFi is not strong, for example, your call could drop. Or, you could accidentally touch the screen and hang up the call. It may be that you just lose your picture or video.

If you lose your picture but you are still in the call or still have audio, the first thing you should do is touch the screen of your device to see if you can bring the picture back. If you can't regain your picture or if you are completely disconnected from the call, you may just have to call the person back (for a call with one person), or rejoin the group call.

There is an important difference here between calls with just one person and group calls. If you get disconnected from a call with just one person, all you have to do is go back into your FaceTime App and call the person you were talking to, again. Group calls are a little different.

If you get disconnected from a group call, press the Home Button to get back to your Home Page and then tap the FaceTime button: 📹 Once you are back in FaceTime, to get back into the group call, tap on the ongoing call that is highlighted in green to join it. You can also join a FaceTime call this way if you missed the notification when it popped up for just a few seconds (see page 15).

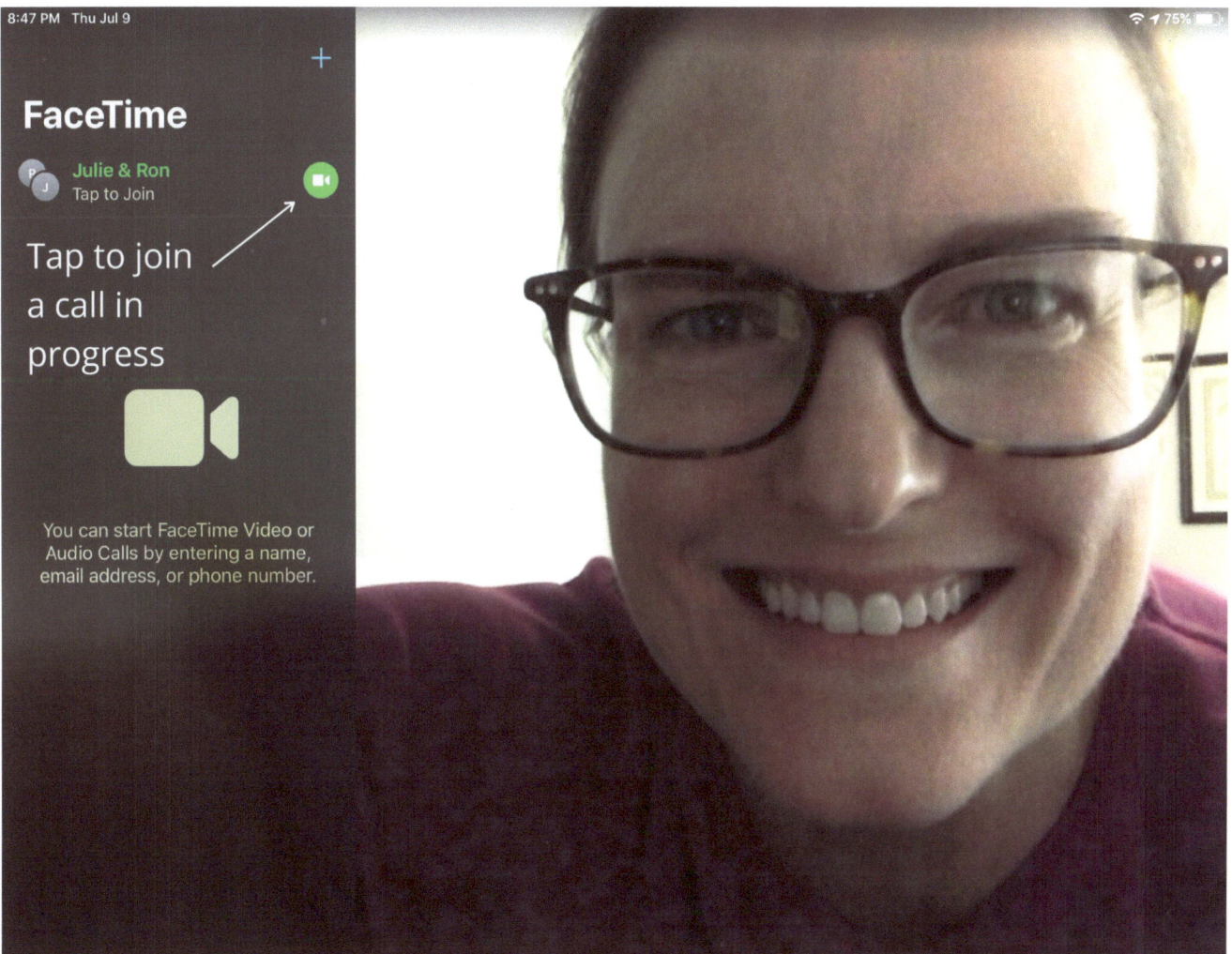

Glossary (words in bold are important to remember):

-**App or Application** - An app is a computer program for your apple device.

-Dropped Call - This is the same as getting disconnected.

-Emoji - A small image or icon that shows an idea, emotion or etc., like a smiley face.

-**Icon** - The image that represents an app, function or tool on your apple device

-iOS - This is the operating system, or computer program that allows your phone to function (like Windows).

-iPadOS - This is the operating system, or computer program that allows your iPad to function.

-Selfie - a picture you take of yourself using your front-facing camera.

-**WiFi** - This is your connection to the internet and it is broadcast through the air, without a wire.

That's it for using your iPad to FaceTime. I hope this guide will help you stay in touch with your family, friends and loved ones during times of separation or across great distances. Just because we can't physically be close to each other doesn't mean we can't stay connected!

If you found this guide useful, please let your friends and family know on Facebook, Instagram and your other social media platforms. If someone bought it for you, please let them know how much you liked it. My goal is to make sure that as many people who need this guide can get access to it.

If you bought this guide for yourself, thank you so much for your purchase. I certainly appreciate it and I hope it helps you stay in touch with your loved ones wherever they are!

If you like this guide, check out my other products about personal finance on my website: newcashview.com.

-Parents' Guide to Better Borrowing - a book about helping parents and high school students figure out how to pay for college with federal student loans.

-Money Basics, Keeping It and Growing It - a book about what I learned about money management as a bankruptcy attorney.

Finally, if you want to see technology videos for seniors when I post them or get notified when more technology for seniors guides become available, type newcashview.com/seniors into your website address bar and sign up today!

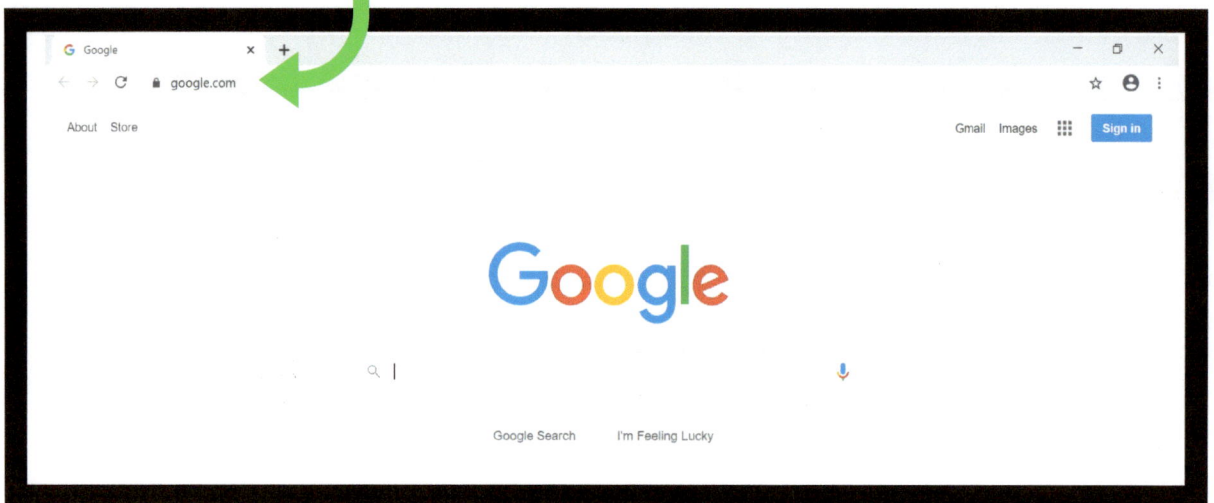

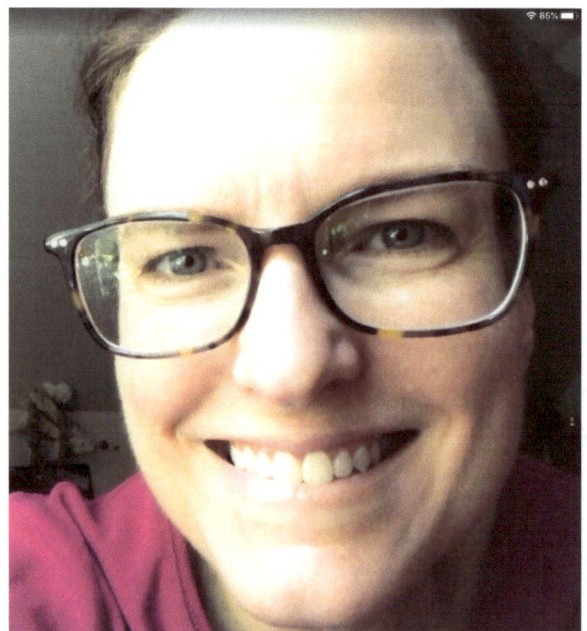

Joy Alford-Brand, JD

Fast Access or "how do I..."

-Make a FaceTime call___11
-Answer a FaceTime call___20
-Take a picture or photo___10
-Add a photo to a contact___9
-Find the cameras on your device___4
-Find the buttons on your device___4
-Get back on a call after getting disconnected___26
-Check to see if I missed any calls___23-25
-Use the effects during a call___16